Breeding Budgerigars How to Establish the Cobalt Shade

By C. Kirby

First Edition February 2017

Published by Wild Budgie Press

© 2017 C. Kirby

This book is copyright. Apart from any fair dealing for the purpose of private study, research, criticism or review, as permitted under the Copyright Act, no part may be reproduced by any process without written permission. Enquiries should be addressed to the Publishers.

All rights reserved.

How to Establish the Cobalt Shade

To many breeders and admirers of the budgerigar, the cobalt shade is the most attractive.

Perhaps that is because it is so difficult to produce, for next to light yellow it is one of the hardest to establish.

Still, the effort is worthwhile.

Pairing for Right Colour

When breeding cobalts it is wise to keep all the dark blues, that is, blue birds bred in the correct line but which are too dark for showing.

Pair these dark birds to deep, evenly-coloured mauves, and usually the resulting cobalts are quite as good in colour as those bred from the more generally favoured cobalt-white mauve mating.

The mating of cobalt to white mauve serves to brighten up colour where brilliance is lacking, but if you can obtain the correct colour and feather quality without having to resort to this, there is no waste of time as in the pairing of splits.

In the breeding of cobalts the best plan is to mate dark blue to mauve, or failing this, cobalt to white mauve.

You will never breed a good class cobalt from the mating of light blue to cobalt, and a few good birds only can be expected from the mating of cobalt to cobalt; cobalt to olive blue will give a proportion of good cobalts.

What the Judge Looks for

A cobalt intended for exhibition should be of first-rate type and of good size, and have prominent throat spots.

This is all very well, of course, but there is a difficulty which has to be overcome. Unless the cobalt carries a deep, even colour all the other features mentioned are of no value in an exhibition.

You may have big birds with very fine necklets, but they may display patchy pale body colour.

Birds of this kind are of value as breeders, but their place is not on the show bench.

When a judge is faced with say, a class of 20 or more cobalts, they select as the winner the bird which carries the highest aggregate of show points, not the biggest bird, not the one of the best type, not the one with the best throat spots, but the bird which has all these features combined with colour.

Photo Gallery

THE END

BIBLIOGRAPHY

1935 'BREEDING BUDGERIGARS.', *The Age (Melbourne, Vic. : 1854 - 1954)*, 6 December, p. 4. , viewed 17 Feb 2017, http://nla.gov.au/nla.news-article203886310

www.ingramcontent.com/pod-product-compliance
Lightning Source LLC
Chambersburg PA
CBHW041120180526
45172CB00001B/354